PERIPLUS NATURE GUIDES

# TROPICAL HERBS & SPICES

Text and recipes by Wendy Hutton

Photography by Alberto Cassio

**PERIPLUS**

**EDITIONS**

Published by Periplus Editions (HK) Ltd.

Copyright © 1997 Periplus Editions (HK) Ltd.
ALL RIGHTS RESERVED
Printed in the Republic of Singapore.
ISBN 962-593-153-8

Publisher: Eric M. Oey
Editor: Kim Inglis
Production: Mary Chia
Additional photography by Luca Invernizzi Tettoni: pages 10, 11,
15, 16, 24 (bottom right), 29, 31, 32, 34, 36 (middle left),
40, 48 (bottom right), 49, 56

**Distributors**

*Indonesia*
C.V. Java Books,
Jalan Kelapa Gading Kirana,
Blok A14 No. 17,
Jakarta 14240

*Japan*
Tuttle Shokai Inc.,
21-13, Seki 1-Chome, Tama-ku,
Kawasaki, Kanagawa 214

*Singapore and Malaysia*
Berkeley Books Pte. Ltd.,
5 Little Road #08-01, Singapore 536983

*The Netherlands*
Nilsson & Lamm B.V.,
Postbus 195, 1380 AD Weesp

*United Kingdom*
GeoCenter U.K. Ltd.,
The Viables Center, Harrow Way,
Basingstoke, Hampshire RG22 4BJ

*United States*
Charles E. Tuttle Co., Inc.,
RRI Box 231-5, North Clarendon,
VT 05759-9700

# Introduction

It is difficult to think of tropical Asia without recalling the heady blend of aromas which permeate the air: the smell of lush vegetation, the faint fragrance of damp soil drenched by a monsoon downpour, a far-off hint of the jungle, the strident notes of strange fruits and above all, the wonderful smells wafting from the region's kitchens.

Perhaps the most distinctive feature of Southeast Asian cuisines is the blending of seasonings—be they fresh herbs, spices or a host of unusual aromatics—to produce some of the world's most flavourful food. In addition, a wide range of sauces, pastes and other manufactured products are part of the cook's repertoire. This book, however, is concerned only with the natural seasonings of the region, both fresh and dried.

During the 19th and early 20th centuries, the countries of Southeast Asia experienced large-scale immigration of Chinese, the majority of them now citizens of their adopted homelands. Similarly, there are also large numbers of Indians now living in Malaysia and Singapore, as well as smaller numbers in cities such as Bangkok and Jakarta. Not surprisingly, both ethnic groups have had an influence on regional cuisines, and some of their herbs, spices and aromatics have been adopted and incorporated in the indigenous cuisines. Although European spices and herbs such as paprika, oregano, rosemary and caraway are sold in many city supermarkets, this book focuses on the seasonings that are used in regional cooking.

For ease of reference, all herbs, spices and aromatics discussed in this book are listed in alphabetical order, without being divided into separate categories. Some plants are used both as a herb and a spice (coriander leaves and coriander seed, for example), while another plant may provide two different spices (nutmeg and mace are from the same fruit).

Most of the spices used regionally will be familiar to cooks around the world. A number are grown in tropical Asia; indeed, some of them—such as cloves and nutmeg—are indigenous to the area; others are imported from China or India. The term "herb" covers a range of leafy plants used for seasoning; apart from several varieties of basil, mint and fresh coriander, there are other herbs such as Indian curry leaves and pandan leaf, lemon grass and the leaves of the Kaffir lime, leaves from the turmeric plant and the pungent polygonum. Aromatics are all the other "vegetable" seasonings which are so important in regional cuisines. Some of these—chillies, garlic and ginger, for example—are known internationally. Others are considerably more esoteric, and include flowers (torch ginger), other rhizomes (turmeric, galangal and Chinese key), sour fruits such as tamarind and nuts which are ground to form part of a seasoning or "curry" paste.

## Buying and Storing Spices

Spices sold throughout Southeast Asia are generally available loose at a spice merchant or market, or packaged in either glass jars (these are most likely imported) or in cellophane bags. As there is usually a considerable turnover of spices, they are generally fresher than those found in the West.

It is important to buy whole spices wherever possible. The flavour of those which have already been ground to a powder will be greatly diminished, unless you buy from a spice merchant who grinds his spices regularly and is likely to have fresh stock.

Modern housewives in a hurry will often buy packaged whole spices. Traditional Asian cooks, however, like to buy their spices whole from a spice merchant, plunging their hand into huge hessian sacks full of coriander or cumin seeds to check they do not contain weevils, scratching at bundles of cassia bark to ensure they're really fragrant. After buying the spices, they pick them over and discard any bits of grit, then wash and sun dry the spices before storing them in glass jars. (Plastic containers permit the volatile oils to escape and should never be used for storage of spices.) Some cooks will then measure out their own blend of spices for a curry powder, and take them to a spice merchant and watch them being ground, ensuring that there is no added rice powder or any other form of adulteration.

Because of the generally humid climate in Southeast Asia, spices will keep fresh in a tightly closed jar or bottle in a cupboard for 1–2 months. However, to keep them almost indefinitely, store the jar or bottle in the deep freeze or normal compartment of a refrigerator.

To maximise the flavour of many spices and to make them easier to grind, heat them gently in a dry pan until they begin to smell fragrant. Take great care not to allow them to change colour, or the flavour will be altered.

## Buying and Storing Herbs and Aromatics

For maximum flavour, cooks in Southeast Asia like to have fresh herbs growing in their garden. This is possible even in an apartment, as many local herbs grow happily in pots. With increasing urbanisation, however, many cooks have to make do with buying herbs in the markets and supermarkets. Most herbs keep for up to 5 days if stored correctly. Any moisture should first be gently wiped off with a cloth, then the herbs wrapped in paper towel or a kitchen cloth (to absorb any condensation) and stored in an airtight glass or plastic container in the refrigerator. Others which should be treated differently are discussed individually under their description.

Many aromatics such as rhizomes, as well as shallots and garlic, can be stored for about a month in a dry, airy place such as a basket, while tamarind pulp keeps almost indefinitely. Other storage tips are included where appropriate.

# Historical Overview of Spices, Herbs and Aromatics

It is perhaps not an exaggeration to say that spices have had a greater impact on the world than any other foodstuff. Fragrant cloves with their woody overtones, heady sweet cardamom, pungent black pepper, the nutmeg whose complex flavours burst forth when grated—all these and other aromatic seasonings of vegetable origin were once so highly prized that some were literally counted out grain by grain.

Spices have been used for thousands of years. The ancient Egyptians, Romans, Greeks, Indians and Chinese all held spices in great esteem, not only for flavouring food and as medicine but also as an ingredient in magic potions, breath sweeteners and to perfume the air. A few spices, such as cumin which is native to Egypt, originated in the Middle East, but the majority grew in Asia, along the Malabar coast of India, in China and in parts of what is now Indonesia.

Both the Chinese and Indians had been trading in parts of Southeast Asia for centuries, and spices which they themselves did not produce were brought to the Asian mainland by sailing boat. From there, camel-laden caravans travelled the Silk Road from western China past northern India and Afghanistan. Their cargoes of silk and spices eventually reached the ports of Syria. Here, they were loaded onto Arab ships for transport to Venice, from where they were distributed to the rest of Europe. By the end of the 14th century, Venice was importing an estimated 180,000 kilograms (around 400,000 lb) of spices annually, around half of it pepper and dried ginger, the rest made up of cloves, cinnamon, mace, cardamom and fresh ginger.

As can be imagined, by the time these spices had travelled as far as the markets of mediaeval Europe, the prices were astronomical and spice cupboards in homes that could afford such luxuries were kept under lock and key. It was in an effort to break the Arabian monopoly on spices provided to the Venetian markets and to therefore reduce their price that European explorers such as Christopher Columbus and Vasco da Gama set out on their epic voyages across uncharted oceans.

India was home to the largest number of spices, especially pepper, cardamom and cinnamon, and their use was described in Sanskrit manuscripts as long as 4,000 years ago. Indian traders voyaging in Southeast Asia, especially to Sumatra and the coast of Indo-China (where their presence was recorded as early as the 5th century AD), not only introduced Hindu beliefs and elements of art and architecture, but also many of their spices.

It is interesting to note that coastal areas of Southeast Asian countries with a long history of contact with India show a much greater usage of spices in their cuisine than inland civilisations which had virtually no direct contact with the outside. One need only compare the richly spiced, complex dishes of

Sumatra (where the Hindu Srivijaya empire was centred) with, for example, the relatively simple food of Central Java. Much more recently, contact by Indian traders in the 19th century (especially in Thailand) and the immigration of Indians to Malaya and Singapore during the same period and well on into the 20th century have had an influence on the use of Indian spices in those countries.

China, even more than India, has influenced the cuisines of the region. Although Chinese trading junks have plied the seas for centuries, it was the immigration of hundreds of thousands of Chinese to every part of Southeast Asia during the past century that had the greatest impact. Ingredients such as soy sauce, beancurd, bean sprouts and noodles were all introduced by the Chinese, as were spices native to China, particularly the cinnamon-scented star anise and cassia.

By a quirk of climate and geography, five tiny islands in the Moluccan archipelago, in what is now Indonesia, were once the sole home of cloves and the nutmeg tree, whose fruit produce both nutmeg and mace. Although the Portuguese reached the area first, the Dutch were the eventual winners in the race for controlling the source of cloves and nutmeg, and in the 17th century, went on to colonise the islands of the Indonesian archipelago, which they named the Dutch East Indies. The Dutch were even more ruthless than the Arab traders had been in their efforts to monopolise the Moluccan spice trade. They restricted the production of cloves to control the prices, and cut down the trees of any grower who had the temerity to plant them without permission. However, clove seedlings were eventually smuggled out of the Moluccas by the 19th century, and now, somewhat ironically, Indonesia imports cloves from Tanzania to meet local demand for their *kretek* cigarettes.

The British had established nutmeg plantations on the island of Penang by the end of the 18th century, later planting them on Singapore (and in the West Indies) as well. The nutmeg has today lost much of its value and supply is greater than the demand. A beautiful tree which produces a golden fruit somewhat like an apricot, the nutmeg inspired a nursery song which implies that the Spanish may have tried to secure some of the trade in this spice:

"I had a little nut tree and nothing would it bear,
But a silver nutmeg and a golden pear;
The king of Spain's daughter came to visit me,
All for the sake of my little nut tree."

In today's world of modern transportation and international trade, spices are treated much the same as any other commodity. They are not only put to culinary use but incorporated in toothpastes, perfumes, cosmetics and soaps. The mystery and rarity of spices have virtually disappeared, but their magical effect on food and their ability to delight the palate remain unchanged.

# Annatto

*BIXA ORELLANA*

**Botanical Family:**
Bixaceae

**Thai name:**
Kam tai

**Malaysian name:**
Jarak belanda, kesumba

**Indonesian name:**
Kesumba

**Tagalog name:**
Achuete, atsuete, echuete

The annatto, a native of tropical America, was introduced to the Philippines by the Spanish. It has spread to other parts of Southeast Asia, where it is a common shrub, grown for its decorative furry red seed pods which look somewhat like an elongated rambutan fruit.

When the annatto pods are ripe, they turn brown and split open to reveal bright red seeds. These are used primarily as a food colouring or dye in the Philippines, but are not used in other Southeast Asian cuisines. Filipino cooks usually fry them gently in a little oil so that the oil takes on a bright red colour. This oil is then used to cook paella and other rice dishes. The seeds can also be soaked in water and the resulting liquid used in recipes where oil would be inappropriate. Today, commercial food colouring has largely replaced the use of annatto seeds.

Annatto is ground and used as a spice in parts of Latin America, although it is not used in this fashion in the Philippines.

# Asam Gelugor

GARCINIA ATROVIRIDIS

This fruit, which is native to Peninsular Malaysia, is a member of the Garcinia family, a family which also includes the highly prized fruit, the mangosteen. The small round fruits of *asam gelugor*, which does not have a common English name, are very sour and therefore not eaten fresh. Instead, they are thinly sliced and dried until shrivelled and brownish black.

*Asam gelugor*, also known as *asam keping* (literally "sour slices"), is used primarily in fish curries in Malaysia and Singapore. Its acidity and flavour are subtly different to the sour fruitiness of the more commonly used souring agent, tamarind, but this can be substituted.

Another member of the same family, a tree known as *goraka* in Southeast India and Sri Lanka, produces a fruit used in similar ways to *asam gelugor*. In Thailand, yet another Garcinia, *G. schomburgkiana*, shares the same sourness as *asam gelugor* and is used fresh in some salads, and also in fish curries. It is known in Thai as *madan*.

**Botanical Family:**
Guttiferae

**Thai name:**
Som khaek, sommawon

**Malaysian name:**
Asam gelugor

**Indonesian name:**
Asam gelugor

# Basil

## OCIMUM BASILICUM, O. AMERICANUM, O. TENNIFLORUM, O. GRATISSIMUM

**Botanical Family:**
Labiatae

**Thai name:**
Bai horapa, bai manglak, bai kaprow

**Malaysian name:**
Daun kemangi, daun selasih

**Indonesian name:**
Daun kemangi, daun selasih

**Tagalog name:**
Balanoi, sulasi

Three varieties of the wonderfully fragrant herb—basil—are found in tropical Asia. Identification by name can be somewhat confusing in Malaysia and Indonesia, as the names *kemangi* and *selasih* are often used interchangeably, or are used to describe one type of basil in one area and another type in a different region. However, the three plants are distinctly different in appearance and flavour.

The closest to European or sweet basil (*O. basilicum*) is the variety most commonly encountered in the region, especially in Thailand and Vietnam. This basil, which is known as *bai horapa* in Thailand, has intensely aromatic dark green leaves and purplish flower heads. It is often eaten raw as a herb in Thailand and Vietnam, served on a platter of fresh herbs which are added to lettuce wrappers for enclosing spring rolls, or nibbled with raw vegetables and a spicy dip. This basil is also added (often by the handful) to stir-fried chicken or beef, or to curries, particularly in Thailand.

Another variety of basil is markedly lemon scented; the leaves are slightly hairy, paler green and smaller than the sweet basil. Lemon basil, which is called *bai manglak* in Thailand, is sometimes fried with seafood in Malaysia and Indonesia. The seed coats of this basil are mucilaginous; Thai cooks soak them in water until the seeds are swollen, then mix them with coconut milk to make a dessert.

"Holy" basil *(kaprow* in Thailand) is sacred to Hindus; this basil has narrower leaves than the two other varieties and is less commonly used in regional cuisines. The herb releases its flavour only when cooked, and is used with fish, chicken and beef.

All three types of basil can be grown easily in the garden or a pot on a verandah simply by putting several stems into the earth. Discard the leaves at the base of the stem, but keep the upper leaves. Water well and, within a few days, the stems will have taken root and will grow vigorously if given sufficient sunshine and water.

# Bilimbi

*AVERRHOA BELIMBI*

**Botanical Family:**
*Oxalidaceae*

**Thai name:**
*Taling pling*

**Malaysian name:**
*Belimbing asam*

**Indonesian name:**
*Belimbing wuluh*

**Tagalog name:**
*Balimbing*

A relative of the popular starfruit *(Averrhoa carambola)*, this small pale green fruit lacks the five ridges which characterise the larger sweet starfruit. Small, smooth and narrow, the bilimibi fruit grows in clusters from the main trunk or branches of a small tree.

As the flavour is highly acidic, bilimbi is cooked to make pickles or sambals, or added to fish curries. Some cooks, especially in Indonesia and Thailand, add slices of the raw fruit to chilli and other ingredients to make a sour and spicy side dish. Although most commonly used in savoury dishes, bilimbi also makes a pleasant jam which turns an attractive pinkish-orange shade.

The skin is very thin, and as the fruit softens and ferments rapidly after ripening, it is not common in big city markets and is more likely to be found in local markets or kitchen gardens. Bilimbi, thought to be a Malaysian native, has a variety of medicinal uses in Southeast Asia, including treatment of fever and skin disorders.

# Indian Borage

*PLECTRANTHUS AMBOINICUS*

This fleshy-leafed herb, thought to be native to India, grows wild in Malaysia. It is easily cultivated and found in pots or planted in kitchen gardens in some other parts of Southeast Asia. Interestingly, it is also found in Australia, where it is popularly known as "five-in-one", and in the West Indies, where it is called broad-leaf thyme.

The leaf is particularly pungent and perhaps closest in flavour to thyme; others think it makes a good substitute for sage. Somewhat confusingly, the Filipinos call it *oregano*, even though it bears no resemblance to that plant, which does not grow in Southeast Asia.

Indian borage is not often used in regional cuisine, although it is sometimes added sparingly to fish or goat meat curries in Malaysia or Java to counteract the strong smell. Just one very finely chopped leaf can be added to a Western-style bread stuffing with a pleasing result. Indian borage is used medicinally as a cough cure, particularly by ethnic Indians.

**Botanical Family:**
*Labiatae*

**Malaysian name:**
*Daun bangun-bangun*

**Indonesian name:**
*Daun kucing, daun kambing*

**Tagalog name:**
*Oregano*

# Candlenut

## ALEURITES MOLUCCANA

**Botanical Family:**
Euphorbiaceae

**Malaysian name:**
Buah keras

**Indonesian name:**
Kemiri

This waxy, beige nut, a native of Malaysia and Indonesia, is related to the Queensland Bush Nut, which is better known internationally as the macadamia. (Perhaps to the chagrin of the Australians, the macadamia, which is an excellent eating nut, was first exploited commercially by the Hawaiians, after they imported the plant from Queensland and developed a way of cracking the hard shell.)

Unlike its relative, the candlenut is never eaten raw as a savoury or dessert nut, but is always cooked. In large quantitites, the candlenut is said to be poisonous, although its purgative qualities, which are strongest when the nut is freshly picked, disappear after it has been kept for a while. The candlenut is used in Malay and Indonesian cuisine. A few nuts are pounded to a paste and used to add texture and flavour to curry-like dishes.

As the candlenut contains a large amount of oil, it can become rancid if kept for any length of time. It is best to store them in a closed container in the refrigerator.

# Cardamom

## ELETTARIA CARDAMOMUM

This richly fragrant spice was known as "grains of Paradise" in mediaeval Europe. It is a member of the ginger family and a native of south India. Introduced to Southeast Asia more than a thousand years ago, plants have been found growing in the ruins of ancient Khmer trading posts.

Cardamom fruits consist of oval capsules or pods, each containing a dozen or so black seeds. The pods are sun dried until pale greenish, although some are bleached so that the pods turn straw white.

Cardamom pods are generally slit, lightly pounded or bruised and added, skin and all, to curries or to rice dishes. The spice also has an affinity with sweet dishes, and the finely ground powder is sometimes sprinkled on the top of Indian milk desserts. The Thais also add cardamom leaves to some dishes of Indian origin.

Another variety of this spice, called black cardamom, has a much larger black pod, but the use of this appears to be confined to India.

**Botanical Family:**
Zingiberaceae

**Thai name:**
Luk grawan

**Malaysian name:**
Buah pelaga

**Indonesian name:**
Kapulaga

# Cassia

## CINNAMOMUM CASSIA

**Botanical Family:**
Lauraceae

**Thai name:**
Ob cheuy

**Malaysian name:**
Kayu manis

**Indonesian name:**
Kayu manis

Somewhat confusingly, the spice commonly sold in Southeast Asia as cinnamon is not true cinnamon (C. *zeylanciaum*) but cassia, from a related species. True cinnamon, the dried bark of a tree native to Sri Lanka, consists of tightly rolled thin quills, pale reddish-brown in colour, with a delicate fragrance and sweet flavour.

True cinnamon is almost never used in Southeast Asia, where the bark of the cassia is now grown. Cassia is native to China, where it is used in some beef dishes. Cassia bark is dark brown, thicker and more strongly flavoured than true cinnamon, and it is also very much cheaper.

Whole pieces of cassia bark are used in some regional cuisines, generally with meat or rice. If powdered cinnamon is required, pieces of bark should be broken and then ground freshly in a coffee mill or spice grinder as commercial cinnamon powder quickly loses its fragrance. Both cassia and cinnamon are used medicinally, while the leaves of the cinnamon tree are used in some Indian dishes.

# Chinese Celery

*APIUM GRAVOLENS*

This somewhat straggly, slender plant bears no resemblance to a similar cultivar, the large white-stemmed celery grown as a vegetable in temperate climates. It is thought that celery is native to both Northern Asia and to Europe, and there is evidence that it was known to the ancient Egyptians, Romans and Greeks.

The celery grown in Southeast Asia is from the Northern Asian cultivar, and its flavour is far more emphatic than the related temperate-climate vegetable.

Chinese celery is rarely used as a vegetable. Only the leaves are used, sparingly, as a herb. They are most commonly added to soups (hence one of its Malay names, "soup leaf"), but a sprig or two may be added to some lentil, stir-fried rice and noodle dishes. Sometimes, a portion of the stems is chopped and added to stir-fried dishes.

Chinese celery is generally sold in markets with other herbs such as spring onions and coriander leaf, rather than being found with leafy green vegetables.

**Botanical Family:**
*Umbelliferae*

**Thai name:**
*Khen chaai*

**Malaysian name:**
*Daun sop, selderi*

**Indonesian name:**
*Selderi*

**Tagalog name:**
*Kintsay*

# Chilli

CAPSICUM SPP.

**Botanical Family:**
Solanaceae

**Thai name:**
Prik kee fah; phrik kee nu

**Malaysian name:**
Lombok, lada, cili, cili padi

**Indonesian name:**
Cabe, lombok; cabe rawit

**Tagalog name:**
Sili; siling labuyo

The chilli is such a hallmark of regional cuisines that it is surprising to reflect that it was unknown in tropical Asia until it was introduced by the Portuguese in the 16th and early 17th centuries. Prior to its availability, pepper was used to provide the pungency or "heat" of regional food.

The chilli is found in a bewildering variety of colours, sizes and flavours in different parts of the world. There are more than two dozen varieties encountered in Southeast Asia, including finger-length chillies in red and green; medium-length plump chillies which can be yellow, pale creamy white, orange, green or red; tiny bird's-eye or, as the Thais call them in an accurate description of their size, "rat droppings" chillies; and short bulbous chillies known as *tabia Bali* and found in that Indonesian island.

The most common chilli is perhaps the finger-length chilli (C. *annum* var. *longum*), of medium intensity on the "heat" scale. This is sold green (unripe), red (ripe) and dried. The flavour and fragrance of green and red chillies differs slightly, and where one particular type is specified in a recipe, this should be used.

When dried, the chilli turns dark reddish brown. Dried chillies are usually cut in 2 cm (¾ inch) lengths and soaked in warm water until softened, then pounded to a paste before being cooked. Some or all of the seeds may be removed according to the desired degree of heat. Dried chillies are commonly used by Malaysian and Sumatran chefs, as they add a deeper red colour to a curry than fresh chillies, and lack the smell. Dried chillies are also dry fried or roasted gently until crisp then ground to a coarse powder and used as a condiment in Thailand.

The tiny fiery hot bird's-eye chilli (C. *frutescens*) is another regional favourite, pounded and added to fresh or raw sambals and side dishes in Thailand. In the Philippines, where local tastes do not run to really pungent foods, bird's-eye chillies are put in a bottle of coconut vinegar which is used as a condiment.

# Chinese Key

*BOESENBERGIA PANDURATA*

**Botanical Family:**
*Zingiberaceae*

**Thai name:**
*Krachai*

**Malaysian name:**
*Temu kunci*

**Indonesian name:**
*Temu kunci*

This rhizome, a member of the ginger family, is probably native to Java and Sumatra. It now grows readily throughout Southeast Asia, from Sri Lanka and India to Indo-China, Malaysia, Indonesia and Thailand.

Chinese key looks like a cluster of long orangey-brown fingers; when the skin is scraped off, the yellow interior is revealed. This herb has an aromatic, spicy flavour. Although its growth is so widespread, it is not widely used as an aromatic, except in Thailand, where it is eaten raw in salads and added to mixed vegetable soups and curries, particularly those made of fish. It can also be made into pickles.

In other regions of Southeast Asia, usage of Chinese key is largely medicinal. The rhizomes are believed to be effective in expelling gas and relieving colic, and are also used to cure diarrhoea and dysentery. The leaves are reputed to be an antidote to certain poisons, although they are not used in cooking.

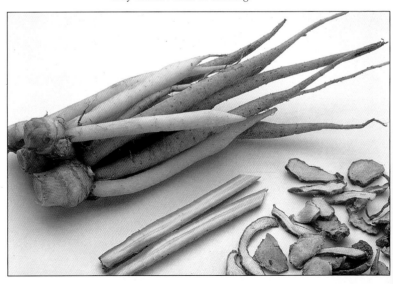

# Chinese Chives

*ALLIUM TUBEROSUM*

This herb has been cultivated in East Asia for centuries, and grows in northeastern India. It is, however, most popular among the Chinese in Southeast Asia. Owing to its rather strong smell, it is also referred to as garlic chives. The chives grow in clusters of long flat leaves (hence another name, flat chives) and are used both as herb and vegetable.

As the flavour of Chinese chives is more pungent than that of spring onions, to which they bear a certain resemblance in terms of length and shape, they are generally lightly cooked rather than eaten raw, adding a distinctive flavour to a number of noodle and other dishes.

Bunches of Chinese chive stems with the flowering culm at the tip are sold and cooked as a vegetable before the flower opens. They are used for their onion flavour. These, like the creamy white Chinese chives sometimes found in the market (the latter grown by cultivating the plant under cover), are considerably more expensive than the regular green chives.

***Botanical Family:***
Lilaceae

***Thai name:***
Bai kuichai, dok kuichai

***Malaysian name:***
Bawang kucai

***Indonesian name:***
Kucai

***Tagalog name:***
Kutsay

# Cloves

## EUGENIA CAROPHYLLUS; SYZYGIUM AROMATICA

*Botanical Family:*
Myrtaceae

*Thai name:*
Garn ploo

*Malaysian name:*
Bunga cingkeh

*Indonesian name:*
Cingkeh

Cloves, indigenous to the Moluccas in Indonesia, were once one of the most sought after spices in the world. They were used by the ancient Egyptians, Romans and Chinese to flavour food, to mask bad breath, as a medicine (especially against tooth decay) and to perfume the air.

The clove is the bud of an attractive medium-sized tree; the buds are picked before they flower and sun dried to a brownish black. Now grown in many other areas than their original home, cloves are often added whole to rice dishes and curries, especially those of Indian origin.

The oil found in cloves (still used by dentists as an antiseptic) is highly volatile, therefore powdered cloves should never be bought. If clove powder is required, heat the cloves gently in a dry pan and then grind them just before using for maximum flavour.

Cloves are used in Indonesia's *kretek* cigarettes, demand being so high that the spice now has to be imported to its original home from Zanzibar and Madagascar.

# Sawtooth Coriander

*ERYNGIUM FOETIDUM*

As the scientific name of this herb indicates, it has a particularly strong smell. This is somewhat reminiscent of true coriander, hence its English name; even the Thai for this herb translates as "foreign coriander".

Sawtooth coriander has been cultivated in Europe since the 17th century, and is thought to have been introduced to Southeast Asia by the Chinese. It is often found growing wild near villages, especially in hill areas.

Both the young and mature leaves of sawtooth coriander are eaten, either raw or cooked. They are often added to strong smelling dishes to mask any unpleasant odours.

Many Asians consider the smell of beef rather pungent, and a few leaves of sawtooth coriander are often added to Thai raw beef salad (*larp*) and beef soups, as well as soups made of offal. A version of the famous Thai *tom yam* soup (*see page 39*) using beef rather than the more common seafood always includes sawtooth coriander.

**Botanical Family:**
Umbelliferae

**Thai name:**
Phak chee farang

**Malaysian name:**
Daun ketumbar jawa

# Coriander

## CORIANDRUM SATIVUM

**Botanical Family:**
*Umbelliferae*

**Thai name:**
*Phak chee*

**Malaysian name:**
*Ketumbar; daun ketumbar*

**Indonesian name:**
*Ketumbar; daun ketumbar*

**Tagalog name:**
*Unsuy*

Coriander leaf is claimed to be the world's most commonly used herb, being indispensable in Southeast Asia, North Asia, India, the Middle East, Mexico and Spanish America.

Often referred to as Chinese parsley in tropical Asia (and as cilantro in the Americas), coriander is actually native to southern Europe and has been used since ancient times. Although fresh coriander is sometimes used in Greek cooking, the dried seeds which are used as a spice are far more commonly encountered in the rest of Europe than the herb. The English name, coriander, comes from the Greek *koris*, which means a bug. (Those not enamoured of the herb claim it smells like bed bugs.)

Raw coriander leaves are widely used as a garnish and flavouring in tropical Asia, and are also pounded to make an excellent fresh chutney or dip in India. It may be reassuring to those who enjoy the distinctive flavour of fresh coriander to know that it is highly nutritious, being rich in calcium, phosphorous, ß-carotene and vitamin C.

Strangely enough, only the Thais seem to have discovered that coriander root also makes an excellent seasoning. In Thailand, it is crushed together with garlic and black pepper to make the most widely used basic seasoning for Thai food.

While the fresh plant is probably the world's most popular herb, the seeds of the coriander can claim to be the most commonly used spice throughout Asia. Coriander seeds are present in almost all blends of curry spices and in a number of other dishes, too. Their faintly orange scent is maximised if the seeds are lightly heated in a dry pan and then crushed to powder just before being cooked.

Fresh coriander grows easily in a pot or garden. Try planting the seeds purchased as a spice (although they may not germinate due to their age, or the fact that they may have been treated). For that reason it may be necessary to buy a packet of seeds intended for planting.

# Cumin

*CUMINUM CYMINUM*

**Botanical Family:**
Umbelliferae

**Thai name:**
Yeera

**Malaysian name:**
Jintan putih

**Indonesian name:**
Jinten

Cumin originated in Egypt, where preserved seeds have been found in the tombs of the Pharaohs. It spread to Greece and then through the rest of Europe, Asia and the Americas, where it is particularly popular in Mexico.

The spice is the small, elongated beige seed of the cumin plant. The flavour is somewhat reminiscent of fennel, dill and aniseed. Cumin closely resembles fennel in appearance, although the latter is whiter, fatter and more strongly aniseed flavoured. Confusion between the two sometimes arises because the Malay names for the cumin and fennel both include the word *jintan*.

In Southeast Asia, cumin is secondary in importance as a spice only to coriander, and is added to a large number of curries, pickles and other dishes. For maximum flavour, buy whole cumin seeds and heat them gently in a dry pan before grinding just before the spice is required.

(So-called black cumin, from Kashmir, is not the same species as true cumin and is not used in regional cuisines.)

# Curry Leaf

*MURRAYA KOENIGII*

The curry leaf tree is not, like the spaghetti tree, a joke but a genuine tree, native to India, which grows wild as far as northern Thailand and is cultivated in Malaysia.

Sprigs from the curry leaf tree contain 12–16 small leaves which have a distinctive fragrance, described by some as resembling anise.

Thanks to the culinary influence of migrant Indians, curry leaves are added, along with other powdered spices, to fish curries prepared by all ethnic groups in Singapore and Malaysia. Freshly ground spice mixtures for fish curry sold in the markets will always contain a sprig or two of curry leaves. The flavour of curry leaf is maximised when fried, and a combination of brown mustard seeds, curry leaves and broken dried red chilli is fried in a little oil and added at the last minute to a number of southern Indian lentil dishes.

The roots, bark and leaves of the curry tree are used medicinally both internally and externally.

**Botanical Family:**
Rutaceae

**Thai name:**
Bai karee

**Malaysian name:**
Daun kari

**Indonesian name:**
Daun kari

# Fennel

*FOENICULUM VULGARE*

**Botanical Family:**
Rutaceae

**Thai name:**
Phong karee, mellet karee

**Malaysian name:**
Jintan manis

**Indonesian name:**
Adas

Only the seeds of the fennel plant are used, as a spice, in tropical Asia, although the plant itself is used as both a herb and a vegetable in Mediterranean Europe. The ancient Greeks used it for slimming, and it is widely acknowledged that both the plant and its seeds have considerable medicinal value.

The fennel plant has leaves which resemble the feathery sprigs of dill. The seeds have a sweet, mellow flavour similar to aniseed, with which it is sometimes confused by culinary writers; aniseed (*Pimpinella anisum*) is not used in the traditional cuisines of Southeast Asia.

Fennel seeds, which look very similar to cumin seeds, contain an essential oil known as anethole, which is used commercially in product flavouring. The seeds are added to the spice mixtures of some curries, particularly fish, and are one of the spices used in *garam masala* (*see page 34*). They are also dry roasted and used as a breath freshner, often handed out in Indian restaurants.

# Fenugreek

*TRIGONELLA FOENUM-GRAECUM*

Fenugreek originated in the eastern Mediterranean, where the plant was used as animal fodder by the Greeks (the botanical name, *foenum-graecum*, means "Greek hay").

Although the leaves are dried and used as a herb in India, in Southeast Asia only the dried seed is encountered. This is used as a spice, the hard, elongated seeds being added whole to some curries (particularly fish), pickles and chutneys.

Fenugreek seeds have a very strong smell described by one herbalist as "malodorous", and should therefore be used sparingly. Excessive amounts are sometimes found in poorly made commercial curry powders in the West, where they tend to overpower all the other spices. (Fenugreek is also a major flavouring of artificial maple syrup.)

Fenugreek seeds are very important medicinally, taken to stimulate lactation in both humans and animals, and are also used for stomach ailments. In Europe in the Middle Ages, the spice was also used as a cure for baldness.

**Botanical Family:**
*Leguminosae*

**Malaysian name:**
*Alba*

# Five-spice Powder

This important Chinese seasoning does not grow on one tree but literally on five, as it is a mixture of five different spices: cassia, cloves, fennel, Sichuan pepper and star anise. The overall fragrance resembles that of aniseed with a hint of cinnamon.

Five-spice powder does not retain its flavour for long, so it is best to buy it in small quantitites and to store it in a glass jar in the refrigerator or deep freeze.

Sometimes referred to by its Cantonese name, *ng heong fan*, five-spice powder is added to a number of simmered or braised dishes, particularly those containing pork. Chinese cooks also make a marinade of five-spice powder, various sauces and ginger juice for fried chicken, or simply rub a mixture of salt and five-spice powder into the skin of chicken before deep frying.

Another popular way of using five-spice powder is to mix it with salt (one part of spice to six parts of salt) as a dip for crisp-fried chicken.

# Galangal

## ALIPINIA GALANGA

A native of Java and the Malay peninsula, this member of the ginger family has a pungency and tang quite unlike that of common ginger (*Zingiber officinale*). It is often referred to as greater galangal, to distinguish it from another variety grown in China, lesser galangal.

The young shoots of the rhizome are pale pink, and are more flavourful and tender than the older beige-coloured rhizomes. Galangal is too spicy to be eaten raw, and is used in slices, chunks or pounded to a paste for various curries and side dishes. When pounding or blending galangal to a paste, first chop it into small pieces as it is often obstinately tough. Perhaps this is why Thai cooks often just bruise a large chunk with the flat side of a cleaver and add it whole to the cooking pot.

Slices of dried galangal are exported from tropical Asia, as are pieces of the young rhizome packed in water inside glass jars. However, nothing quite matches the inimitable jungle fragrance of fresh galangal.

**Botanical Family:**
Zingiberaceae

**Thai name:**
Kha

**Malaysian name:**
Lengkuas

**Indonesian name:**
Laos

# Garam Masala

This spice is a mixture of ground spices used in Indian cooking in Southeast Asia—and, of course, in India itself. There is no single recipe for *garam masala*, as this varies according to taste and region.

*Garam masala* is not like a normal "curry powder" or mixture of ground spices, used to season food at the beginning of cooking. Instead it is added as an additional seasoning, sprinkled over many dishes at the end of cooking to give extra fragrance and flavour.

The overall flavour of *garam masala* is generally sweet. A basic combination might include 2 tablespoons of coriander seed, 1 tablespoon cumin seed, 2 teaspoons black peppercorns, 1 teaspoon cardamom seeds (removed from the pods), 1 teaspoon of cloves, a 7.5 cm (3 inch) stick of cinnamon and a large chunk of whole nutmeg. All spices except the nutmeg should be lightly roasted separately in a dry pan, then everything ground to a fine powder. The mixture is bottled only when cool.

# Garlic

*ALLIUM SATIVUM*

Garlic is a member of the same botanical family as onions, leeks and spring onions. It probably originated in central Asia, but its usage is now worldwide.

More than one type of garlic is sold in Southeast Asia; one variety, the so-called Mexican or Italian type, has a pinkish purple tinge to the skin and tends to be richer in flavour and less dry than the white-skinned variety. In tropical Asia, locally grown garlic cloves are often only half the size of their Western counterparts. Garlic is often used in large amounts regionally, particularly in Thailand. It is commonly pounded as part of a seasoning paste; chopped and added to stir-fry dishes and stews; sliced and deep fried until crisp and used as a savoury garnish for vegetables, soups and noodles.

Garlic has no significant nutritive value, but is important medicinally, containing an antibiotic substance which inhibits the growth of some bacteria and fungi. Garlic also helps reduce blood pressure and cholesterol levels.

**Botanical Family:**
Lillaceae

**Thai name:**
Krathiem

**Malaysian name:**
Bawang putih

**Indonesian name:**
Bawang putih

**Tagalog name:**
Bawang

# Ginger

## ZINGIBER OFFICINALE

Around 400 members of the ginger family grow wild in tropical Asia, but this particular variety is the one universally known as ginger. Two forms of the common edible ginger are found in Asian markets: young ginger has very thin skin, is pale yellow and has pinkish shoots with green stalk ends, while old ginger is beige-brown with a thicker skin. This should be scraped off with a knife before using.

*Botanical Family:*
Zingiberaceae

*Thai name:*
Khing

*Malaysian name:*
Halia

*Indonesian name:*
Jahe

*Tagalog name:*
Luya

Young ginger is more tender and juicy than the mature rhizome, so it is preferred for grating or pounding to extract the juice, a popular marinade with Chinese chefs. It can be eaten raw, and is also pickled (a treatment very popular in China and Japan).

Mature ginger, although sometimes served raw in very fine shreds, is more commonly cooked as the flavour is more emphatic than that of young ginger.

Ginger is widely used for medicinal purposes throughout Asia, particularly to improve digestion and to counteract nausea and vomiting.

# Kencur

## KAEMPFERIA GALANGA

This plant is subject to a lot of confusion, exacerbated perhaps, by its lack of a common English name. Some people refer to it as zedoary, although this is actually *Curcuma zedoaria*. Others confuse it with lesser galangal, a completely different rhizome found in southern China.

Although known to some botanists as Resurrection Lily, this camphor-scented rhizome is known only by its local name in the countries where it is eaten: *kencur* (once spelled *kentjoer*) in Indonesia; *cekur* in Malaysia and *pro hom* in Thailand.

The white or yellowish rhizome (there appear to be two varieites grown) is pounded and added to a number of dishes, and is particularly popular in Bali. In Thailand, both the rhizomes and leaves are added to fish curries, and the young leaves are also eaten as a raw vegetable with a spicy shrimp paste sauce, *kapi kua*. In Malaysia, the tightly packed leaves, which grow very closely to the ground, are finely sliced and used as a herb in salads.

**Botanical Family:**
Zingiberaceae

**Thai name:**
Pro hom

**Malaysian name:**
Cekur

**Indonesian name:**
Kencur

# Lemon Grass

## CYMBOPOGON CITRATUS

**Botanical Family:**
Gramineae

**Thai name:**
Takrai

**Malaysian name:**
Serai

**Indonesian name:**
Sereh

**Tagalog name:**
Tanglad

The distinctive lemony fragrance of this grass, which is related to citronella, is a hallmark of much Southeast Asian cuisine. Lemon grass grows readily in almost any soil, its bulbs and leaves constantly multiplying. It is easily grown by putting the stem end of a stalk of lemon grass in water until the roots appear, then it can be potted or put out in the garden.

The coarse, long flat leaves are normally discarded, and only around 10–15 cm (4–6 inches) of the bulbous base used. If lemon grass is to be eaten raw, the outer layers of the bulb should be peeled away until the pinkish ring inside appears; this tender portion is then finely sliced. Lemon grass is also bruised and added whole to many curry dishes, or sliced before being pounded to a paste.

Whole stalks of lightly bruised lemon grass, trimmed to a length of about 15–20 cm (6–8 inches) make an excellent skewer for prawns or seafood satay, as is sometimes enjoyed in Bali.

# Kaffir Lime Leaf and Zest

## CITRUS HYSTRIX

The Kaffir lime has a dark green, warty skin which inspires an alternative name, leprous lime. There is virtually no juice inside, but the lime is valued for its zest and also for the marvellous perfume of its double leaf—no other lime or lemon or citrus can match it.

Grated Kaffir lime rind is added to some dishes, while the distinctive leaves are also used as a herb, particularly in Thailand (a spicy *tom yam* soup [*see page 25*] would be unthinkable without these).

Kaffir lime leaves are very finely shredded and added raw to some salads, or added to cooked food. They are often used by the Nonya cooks of Malaysia and Singapore, whose cuisine makes liberal use of fresh herbs.

It is sometimes difficult to find a regular supply of kaffir lime leaves in the markets; when you are able to obtain them, buy a large amount and store them in a sealed plastic bag in the deep freeze; they retain all their flavour and texture after a brief thawing out.

**Botanical Family:**
*Rutaceae*

**Thai name:**
*Bai magrut*

**Malaysian name:**
*Daun limau purut*

**Indonesian name:**
*Daun jeruk purut*

# Melinjo

## GNETUM GNEMON

**Botanical Family:**
Gnetaceae

**Thai name:**
Pee sae

**Malaysian name:**
Daun Melinjo

**Indonesian name:**
Daun Melinjo

A native of Southeast Asia, the *melinjo* tree is found wild in the forests and is also cultivated in some areas. The leaves are treated as a herb and a vegetable, although their use is very localised. There does not appear to be a common English name for this tree, known as *melinjo* in both Malaysia and Indonesia. One botanist, however, refers to it as Spinach Joint Fir, certainly a name no English-speaking Asian will have ever heard of.

Thai cooks use the young shoots and infloresences either raw or cooked, and add the young leaves to mixed vegetable soup. Although the tree can be found in Malaysia, very few local cooks seem to make use of it and one has to go to Indonesia to find it readily in the markets.

Sprays of young leaves and immature nuts with a fleshy green to red coating are an essential ingredient in the West Javanese sour vegetable dish, *Sayur Asam*. The mature nuts are dried and pounded to make a very popular *emping* or deep-fried wafer.

# Mint

*MENTHA ARVENSIS*

Mint is a temperate climate herb more commonly associated with Western cooking. A number of varieties are found in Europe, including peppermint, spearmint, apple mint and even eau de cologne mint.

The variety cultivated in parts of Southeast Asia is a spearmint. This grows easily in the highlands, but can also be grown in lowland areas in the shade, providing it receives adequate water.

The use of mint seems to be restricted primarily to Thai, Vietnamese, Laotian and Indian cooks; although a variety grows in Java (*M. javanica*), it is almost never used there. Sprigs of mint, which have a particularly refreshing tang compared with some other herbs, are added liberally to many salads in Thailand. Mint sprigs are used as a garnish for some Indian dishes, and made into an excellent fresh mint chutney. Some Malaysian and Singaporean cooks add mint to the spicy noodle soup, *laksa*, although others prefer the polygonum (*see page 47*).

**Botanical Family:**
Labiatae

**Thai name:**
Bai saranae

**Malaysian name:**
Daun pudina

**Indonesian name:**
Pudina

# Mustard Seed (Brown)

BRASSICA JUNCEA

**Botanical Family:**
Cruciferae

**Thai name:**
Mustard

**Malaysian name:**
Biji sawi

**Indonesian name:**
Biji sawi

Three types of mustard plant produce seeds. The very pungent black mustard (*Brassica nigra*) is native to Europe and was traditionally used in making the condiment, mustard. The very best French mustard from Dijon still uses this variety. Another European variety (*Sinapis alba*) produces yellowish-white seeds of a much milder flavour than black mustard, and is used to make American mustard, as well as being used in pickles.

Brown mustard produces seeds that are similar to black mustard but reddish-brown in colour. These seeds are widely used in Sri Lanka and India, and in Malaysia and Singapore (due to the Indian immigrant influence).

The rather coarse taste of brown mustard is changed to a pleasant nutty flavour after frying, so mustard seeds are almost always cooked in a little oil until they start crackling. They are then added to *dal* and vegetable combinations. The seeds are also ground and added to spice pastes, pickles and chutneys.

# Nutmeg

*MYRISTICA FRAGRANS*

Like the clove, this spice is native to the Moluccas and has been used in the eastern Mediterranean since the 12th century, when it was taken there by Arab traders.

The nutmeg tree, now grown in many areas outside its original home, produces fruit with firm yellow flesh. This flesh is pickled or soaked with sugar syrup and eaten as a confectionery or an antidote to seasickness in Asia.

Inside the fruit is a nut, the nutmeg, covered by a hard shiny brown shell. On the outside of this shell is a vivid red lacey web or aril. This is mace (pictured below left), which is also used as a spice in the West but which is used only in dishes of Indian origin in Southeast Asia.

The dried nutmeg will keep almost indefinitely, and should be grated or crushed just before using. Often nutmegs are sold in Asia still with their hard shell, which should be broken off and discarded. Although nutmeg is a popular sweet flavouring in the West, it is used only in savoury dishes in Southeast Asia, such as curries and soups.

**Botanical Family:**
Myristicaceae

**Thai name:**
Chan thet

**Malaysian name:**
Buah pala

**Indonesian name:**
Pala

# Pandan

## *PANDANUS AMARYLLIFOLIUS*

**Botanical Family:**
Pandanaceae

**Thai name:**
Bai toey hom

**Malaysian name:**
Daun pandan

**Indonesian name:**
Daun pandan

**Tagalog name:**
Pandan

The earthy fragrance of this member of the pandanus or screwpine family gives a distinctive flavour and aroma to many dishes throughout tropical Asia.

Consisting of clumps of long, blade-like leaves, pandan grows easily, especially in damp areas on a slope and is found in many kitchen gardens.

The pandan leaf is remarkably versatile. Throughout the region, a leaf or two is added to rice before cooking to give it the fragrance of newly harvested grain. It is also used to give colour and flavour to desserts and cakes; the leaf is either raked with the tines of a fork to release its flavour, or pounded to extract the green juice. Modern cooks put the coarsely chopped leaves in a blender with a little water to obtain the juice; some even buy artificial pandan essence sold in bottles like vanilla.

In Thailand, pandan leaves are fashioned into containers for food, or wrapped around morsels of marinated food before frying.

# Pepper

*PIPER NIGRUM*

Pepper, once such a valuable spice that it was literally sold by the grain in mediaeval Europe, comes from the dried berries of a vine native to the Malabar coast of India.

The pepper vine, now grown in much of tropical Asia—particularly in the East Malaysian state of Sarawak—produces black, white and green peppercorns. The black variety, which are the whole berries picked unripe and sun dried until shrivelled, is the most widely used form of pepper in Southeast Asia. Before the arrival of chillies from the Americas in the 16th century, peppercorns were the main source of heat in tropical Asian food.

White peppercorns are ripe berries with the red skins removed before being bleached white by drying in the sun. Chinese cooks are the main users of white pepper, sprinkling it in powdered form on many cooked dishes.

Green peppercorns, which are not used by Southeast Asian cooks, are the immature berries pickled in brine or freeze-dried while still fresh.

**Botanical Family:**
*Piperaceae*

**Thai name:**
*Phrik Thai*

**Malaysian name:**
*Lada hitam*

**Indonesian name:**
*Merica*

# Pepper (Wild)

*PIPER SARMENTOSUM*

**Botanical Family:**
Piperaceae

**Thai name:**
Chaa phluu

**Malaysian name:**
Daun kaduk

**Indonesian name:**
Kadok

This shade-loving plant grows wild in the dry, evergreen forests of Thailand and Vietnam, and is cultivated in Malaysia and Indonesia. The shiny, heart-shaped leaves of the wild pepper (known as *la lot* by the Vietnamese), are eaten raw when young, or can be cooked.

In a favourite snack in Thailand, *mieng kum*, wild pepper leaves are used as wrappers for an assortment of finely chopped pieces of raw ginger, peanuts, dried shrimps, shallots and lime. In Peninsular Malaysia, particularly in the northern states near the Thai border, wild pepper is one of the herbs which are finely sliced and used in the herb and rice salad known as *nasi ulam* or *nasi kerabu*.

The whole wild pepper plant is believed to have medicinal value. The Chinese and Thais crush wild pepper roots with salt to relieve toothache, while in Indonesia, they are chewed with betel nut and the resulting juice swallowed as a remedy for coughs and asthma. Malaysians use the leaves for headaches and pains in the bones.

# Polygonum

*POLYGONUM ODORATUM*

A native of Southeast Asia, this plant grows wild in ditches or on the banks of streams and ponds. Polygonum does not seem to have a widespread English name, although Western herbalists refer to it as a species of knotweed. More commonly used terms in Asia are laksa leaf and Vietnamese mint (it is widely used in Vietnam).

The long, arrow-shaped leaves are one of the most strongly flavoured herbs used in Southeast Asia. Many Chinese in Singapore and Malaysia do not consider *laksa*, the spicy noodle dish, complete without chopped polygonum on top (although in Sarawak, the famous *Kuching laksa* is always topped with fresh coriander). Sprigs of polygonum are served with other fresh herbs and lettuce leaves to wrap Vietnamese spring rolls, and are eaten raw with spicy dips in Thailand.

Polygonum grows easily in a pot; keep some stems bought in the market in a jar of water until the roots start to sprout, then plant and keep well watered.

**Botanical Family:**
Polygonaceae

**Thai name:**
Phak phai

**Malaysian name:**
Daun laksa, daun kesum

# Poppy Seed

## PAPAVER SOMNIFERUM

**Botanical Family:**
Papaveraceae

**Malaysian name:**
Kas kas

The poppy seed used in cooking comes from the opium poppy. The plants have been cultivated in China and India for almost 2,000 years, and used medicinally as a pain reliever. Opium is produced from the latex extracted from the green seed pod of the poppy plant. By the time the poppy seeds are ripe, all traces of the opium are gone and the seeds have no narcotic quality whatsoever. (This fact has not, however, prevented the Singapore government from banning their sale for culinary use).

Although black poppy seeds are used widely in Europe, the variety used in Southeast Asia comes from India and has smaller, pale creamy white seeds.

White poppy seeds are usually ground to a paste and used to thicken a number of curries of Indian origin. They have very little flavour unless previously dry fried until pale golden, which brings out a pleasant nutty flavour. Poppy seeds are sometimes used in Malay-style curries, and are popular with ethnic Indian cooks.

# Salam Leaf

## SYZYGIUM POLYANTHA

This large, dark green leaf is a very popular herb in Indonesia, and although the tree (a member of the cassia family) will grow in other parts of Southeast Asia, only Indonesian cooks seem to use it.

Sprigs of fresh or even slightly dry salam leaves, usually measuring 8–10 cm (3–4 inches) in length, are sold in most markets. It is also possible to buy packets of the dried leaf, which still releases its aroma when cooked. The young leaves at the tips of the stem are lighter in colour than the mature leaves.

Salam leaf is not used raw, but is added—usually whole—to various curry-like dishes, stews and some sauces. It is difficult to describe the flavour, and even more difficult to recommend a substitute for this herb. A number of cook books translate *daun salam* as bay leaf, to which it bears no resemblance whatsoever in flavour. Rather than using bay leaf as a substitute if salam leaf is unavailable, omit the herb altogether.

**Botanical Family:** Myrtaceae

**Malaysian name:** Daun salam

**Indonesian name:** Daun salam

# Sesame Seed

*SESAMUM INDICUM*

**Botanical Family:**
Pedaliaceae

**Thai name:**
Ngaa

**Malaysian name:**
Bijan

**Indonesian name:**
Wijen

**Tagalog name:**
Linga

The sesame plant is thought to have originated in Africa, but has been cultivated in China and India for centuries. The seeds are small, flat and pear-shaped. The colour is generally creamy white, although black sesame seeds (usually from China) are also used in Southeast Asia. Sesame seeds are often used as a coating, to give a delicate nutty flavour to either savoury or sweet foods.

Indian gingelly oil, light and almost flavourless, is pressed from sesame seeds, which contain about 50% oil, and is used for frying, pickles and chutneys.

Both Chinese sesame oil, which is used only as a seasoning and not as a frying medium, and Chinese sesame paste are made from toasted sesame seeds. So, too, is the popular sesame paste which is mixed with cooked vegetables in some Japanese recipes. East Asian sesame oils and pastes are, because of the initial frying, much stronger in flavour than Indian gingelly oil or Middle Eastern sesame paste (*tahini*), where raw sesame seeds are used.

# Shallot

*ALLIUM ASCOLONICUM*

Shallots grown in Southeast Asia have a purplish skin which encloses a delicately flavoured interior. They grow in clusters, similar to the much larger brown-skinned European shallot.

Their aromatic flavour makes them an important addition to many seasoning pastes. *Rempah* in Malaysia and *bumbu* in Indonesia, for example, use shallots along with other herbs, roots or rhizomes and dried spices. Shallots are preferred to onions not only for their better flavour but because they contain less moisture. Spice pastes can thus be fried rather than stewed in oil at the beginning of cooking, a subtle, but important detail.

**Botanical Family:**
Lillaceae

**Thai name:**
Horm lek, horm daeng

**Malaysian name:**
Bawang merah

**Indonesian name:**
Bawang merah

**Tagalog name:**
Sibuyas tagalog

Shallots are also treated as a seasoning in pickles and salads throughout the region, their mild flavour making them perfectly palatable and easily digestible while still raw.

Slices of deep-fried shallot are one of the most widely used garnishes in Malay and Indonesian cooking, and are sold pre-cooked in plastic bags in markets.

# Sichuan Pepper

## ZANTHOXYLUM PIPERITUM

**Botanical Family:**
Rutaceae

**Thai name:**
Ma lar

This spice is not a true pepper, but is the dried berry of the fagara or prickly ash tree, closely related to the prickly ash found in North America.

Sometimes referred to as Chinese pepper, Sichuan pepper (as its name suggests) is particularly popular in the western province of Sichuan, where it adds a pungent flavour to many dishes. It is also grown in Japan, where it is known as *sansho* or, not surprisingly, as Japanese pepper.

The reddish brown berries should be gently heated in a pan until crisp, then crushed. Like five-spice powder, Sichuan pepper is commonly combined with salt (1 part of crushed pepper to 6 parts of salt) to make a dip which is excellent with fried chicken.

Known as *hua jiao* in Mandarin Chinese, this spice is normally sold by Chinese stores; the powdered version packed in Japan, *sansho*, is sold in small bottles in Japanese supermarkets. The principal effect of this "pepper" is that it numbs the lining of the mouth when eaten.

# Spring Onion

## ALLIUM FISTULOSUM

The spring onion is widely used throughout Southeast Asia, as well as in the north in China, Korea and Japan. It is used raw as a herb, to give flavour and a touch of bright green colour to many soups, noodle and other dishes. Spring onion is also an important garnish, as it can easily be carved into decorative tassles or "brushes".

Spring onions grow in clusters, generally of two to five plants. They have hollow round green leaves tapering to a white stem, which is often covered at the base with an outer layer of purplish skin.

This plant is known by a wide variety of names in Western countries: Welsh onions, Japanese bunching onions, scallions and even as shallots (even though they bear no resemblance to shallots). The green onion, looking like a fatter version of the spring onion, belongs to another variety (*A. cepa*) and has a more emphatic flavour.

Spring onions keep well if refrigerated with the stem ends placed in a glass of water.

**Botanical Family:**
Lillaceae

**Thai name:**
Ton horm

**Malaysian name:**
Daun bawang

**Indonesian name:**
Daun bawang

**Tagalog name:**
Sibuyas-na-mura

# Star Anise

*ILLICIUM VERUM*

**Botanical Family:**
*Illiciaceae*

**Thai name:**
*Poy kak, dok chan*

**Malay name:**
*Bunga lawang*

**Indonesian name:**
*Bunga lawang*

**Tagalog name:**
*Sanque*

This Chinese spice is considerably more pungent than the similarly flavoured European spice, anise, just as Chinese cassia is like a stronger version of cinnamon. In fact, star anise contains the same essential oil, anethole, as the true anise and fennel.

Star anise is the dried flower head of a tree which is a member of the magnolia family. Due to its Chinese origins, it is popular with Chinese, who tend to use it more liberally than other cooks. Star anise is, however, found in most countries in the region, and is often encountered in Thailand, Malaysia and Sumatra.

Star anise looks like an eight-pointed star, each point containing a shiny brownish black seed. The flower is normally added to dishes whole. Sometimes, however, just a few "points" of the star anise are called for in recipes. It is not, unlike other spices, ground to a powder, and the pieces of star anise are simply removed from the curry or gravy when the dish is served.

# Tamarind

*TAMARINDUS INDICA*

The tamarind tree is very tall and graceful, with sprays of fine leaves. The young sprigs and the flowers are used as a vegetable, particularly in Thailand, and also in the Philippines where they are added to soup to give a sour tang. However, it is the fruit of the tamarind tree which is widely used as a flavouring in Southeast Asia.

The long pods of the tamarind contain a number of pulp-covered seeds. When these are young and green, they are added whole to give sourness to the popular Sinigang soups of the Philippines. When the fruits are ripe, they can be eaten fresh with salt and chilli or a dip, a popular Thai approach, or coated with sugar to make a candy, particularly in the Philippines.

Most commonly, however, tamarind fruits are dried and sold as a pulp in packets. A tablespoonful or two of tamarind pulp is soaked for about five minutes in warm water. The juice is strained to remove fibres and seeds, and then used to add a fragrant, fruity sourness to many dishes.

**Botanical Family:**
Leguminosae

**Thai name:**
Ma khaam

**Malaysian name:**
Asam jawa

**Indonesian name:**
Asam jawa

**Tagalog name:**
Sampalok

# Torch Ginger

*ETLINGERA ELATIOR*

**Botanical Family:**
Zingiberaceae

**Thai name:**
Kaalaa

**Malaysian name:**
Bunga kantan, bunga siantan

**Indonesian name:**
Combrang

The pale pink bud of this striking wild ginger, which grows as much as five metres (16 feet) tall, has an intriguing fragrance and slight tang. Although believed to be a native of Java and now grown throughout tropical Asia from Sri Lanka as far as New Guinea, it is rarely found in Javanese cuisine. In fact, its usage seems to be restricted mainly to Thailand, Malaysia and Singapore.

The torch ginger bud opens to form a particularly beautiful flower, but for culinary purposes, it is picked while the bud is still tightly folded. The bud is eaten either raw or treated as an aromatic in cooking. It is halved lengthwise and added to some fish curries, especially those prepared by Peranakan cooks in Singapore and Malaysia. It is also finely sliced and added to some salads, including the popular *rujak* of the same area.

In Thailand, the buds are eaten raw with a spicy dip and generally added to the popular southern salad known as *khaao yam*.

# Turmeric

*CURCUMA DOMESTICA*

A member of the prolific ginger family, turmeric is cultivated for its flavour and vivid yellow colour. In India, the rhizomes are dried and crushed to form powdered turmeric, but in Southeast Asia, the fresh rhizome is generally preferred.

The juice extracted from crushed turmeric is favoured for giving a bright yellow colour to ceremonial rice dishes in Southeast Asia. It was also widely used as a dye for cloth, but has been replaced these days by commercial dyes.

As the flavour and colour of fresh turmeric rhizomes is so intense, it is used in small quantities, often no larger than the size of a pea. Care should be taken not to stain clothing or utensils when using fresh turmeric.

Fresh turmeric leaves are used as a herb in some Malay and Indonesia dishes. In Thailand, young shoots and inflorescences are boiled as a vegetable. The inflorescences can also be cooked with eggs to make an unusual omelette. The rhizome is used in folk medicine throughout the region.

***Botanical Family:***
Zingiberaceae

***Thai name:***
Khamin

***Malaysian name:***
Kunyit

***Indonesian name:***
Kunyit

***Tagalog name:***
Dilaw

# Coriander/Mint Chutney

*A fresh Indian chutney that goes well with roasted or grilled chicken or lamb, this can also be used as a dip. A spoonful or two added to plain yoghurt or to a basic vinaigrette makes an unusual cross-cultural dressing for salads.*

- 1 cup firmly packed fresh coriander or mint leaves
- 8 shallots or 1 medium red or brown onion
- 1–2 fresh green chillies (some seeds removed if preferred)
- 1 slice fresh ginger
- 1$\frac{1}{2}$ tablespoons freshly grated coconut, or desiccated coconut moistened with a little water
- 1 teaspoon salt
- 1 teaspoon sugar
- 1 heaped tablespoon tamarind pulp
- $\frac{1}{4}$ cup warm water

Wash, dry and coarsely chop the coriander or mint. Slice or chop the shallots, chillies and ginger and put together with the coriander or mint in a blender. Add the coconut, salt and sugar.

Soak the tamarind pulp in warm water for 5 minutes, then squeeze and strain to obtain the juice. Add the juice to the blender and process everything until fine. Cover and refrigerate until needed.

# Curry Powder

*Cooks in Asia do not use one blend of spices for all dishes, but change the mixture depending on the basic ingredient to be cooked. Below are spice mixtures for both meat and fish. The method of preparation and storage is the same.*

Meat Curry Powder
300 g (10 oz) coriander
75 g (2½ oz) cumin
75 g (2½ oz) fennel
75 g (2½ oz) dried red chillies
45 g (1½ oz) black peppercorns
45 g (1½ oz) turmeric powder
15 g (½ oz) cinnamon or cassia sticks
10 whole cardamom pods, slit
10 whole cloves

Fish Curry Powder
300 g (10 oz) coriander
75 g (2½ oz) fennel
75 g (2½ oz) dried red chillies
45 g (1½ oz) cumin
45 g (1½ oz) turmeric powder
30 g (1 oz) fenugreek
30 g (1 oz) black peppercorns

Pick over the spice seeds and discard any grit. Heat a large wok and very gently fry the spices (except turmeric powder), one by one, until fragrant. Allow to cool slightly then grind all the spices in a coffee grinder until fine. Sieve into a bowl and add turmeric powder, stirring to mix all spices thoroughly. Bottle when completely cold.

# Sambal Blacan

*This freshly made chilli condiment or* sambal *is the universal favourite in Malaysia and Singapore, with variations also found in Thailand, where it is known as* nam phrik kapi. *The distinctive flavour of the* sambal *is provided by pungent dried shrimp paste, which must be well cooked before being ground with fresh chillies. The final touch is a squeeze of juice from the musk lime or kalamansi, known in Malaysia and Singapore as* limau kesturi. *Ordinary lime or even lemon juice could be substituted, but the* sambal *will lack the distinctive fragrance and flavour of the musk lime.*

> 3 teaspoons dried shrimp paste
> 5–6 fresh red chillies (finger length)
> pinch of salt
> 2–3 musk limes or kalamansi

Press the dried shrimp paste with the back of a spoon to make a thin flat cake. Either place it under a hot grill, or put it into a dry pan and cook over moderate heat, turning it over, until it is dry on both sides and crumbly.

Slice the chillies, removing some of the seeds if you do not want the *sambal* to be too hot, and pound them in a mortar with salt until fine. Alternatively, process the chillies in a blender. Add the cooked shrimp paste and mix well. Serve in small dishes with the cut limes, allowing each person to squeeze in the juice to taste.

# Tandoor Spice Powder

*Food cooked in a* tandoor *or clay oven with hot coals in the bottom of it is originally from northwest India, but this style of cooking is now found in restaurants in most parts of India, as well as in many cities in Southeast Asia.*

- 4 teaspoons turmeric powder
- 2 teaspoons **garam masala**
- 1 teaspoon chilli powder
- 1/4 teaspoon white pepper powder
- 1 teaspoon freshly ground cardamom
- 1 teaspoon garlic powder
- 1/4 teaspoon red food colouring powder or 2 teaspoons paprika

Combine all the ingredients, mixing well, and store in a glass bottle. This mixture of spices can be combined with plain yoghurt, some pounded fresh coriander leaves, a little crushed fresh ginger and salt to make a marinade for chicken or fish. To obtain an authentic flavour, the chicken or fish should be marinated for at least 8 hours and then baked in a *tandoor*, but roasting it in an oven or cooking it over a barbecue also provides excellent results.

# Index by Scientic Name

*Aleurites moluccana* . . . . . . . . . . . . . . . 16
*Alipinia galanga* . . . . . . . . . . . . . . . . . 33
*Allium ascolonicum* . . . . . . . . . . . . . . 51
*Allium cepa* . . . . . . . . . . . . . . . . . . . . . 53
*Allium fistulosum* . . . . . . . . . . . . . . . . 53
*Allium sativum* . . . . . . . . . . . . . . . . . 35
*Allium tuberosum* . . . . . . . . . . . . . . . 23
*Apium gravolens* . . . . . . . . . . . . . . . . 19
*Averrhoa belimbi* . . . . . . . . . . . . . . . 14
*Averrhoa carambola* . . . . . . . . . . . . . 14
*Bixa orellana* . . . . . . . . . . . . . . . . . . . 10
*Boesenbergia pandurata* . . . . . . . . . . 22
*Brassica juncea* . . . . . . . . . . . . . . . . . 42
*Capsicum* spp. . . . . . . . . . . . . . . . . . . 20
*Capsicum annum* var. *longum* . . . . . . 20
*Capsicum frutescens* . . . . . . . . . . . . . 20
*Cinnamomum cassia* . . . . . . . . . . . . 18
*Cinnamomum zeylanciaum* . . . . . . . 18
*Citrus hystrix* . . . . . . . . . . . . . . . . . . .39
*Coleus amboinicus* . . . . . . . . . . . . . . 13
*Coriandrum sativum* . . . . . . . . . . . . 26
*Cunimum cyminum* . . . . . . . . . . . . . 28
*Curcuma domestica* . . . . . . . . . . . . .57
*Curcuma zedoaria* . . . . . . . . . . . . . .37
*Cymbopogon citratus* . . . . . . . . . . . . .38
*Elettaria cardamomum* . . . . . . . . . . 17
*Eryngium foetidum* . . . . . . . . . . . . . . 25
*Etlingera elatior* . . . . . . . . . . . . . . . . 56
*Eugenia carophyllus* . . . . . . . . . . . . . 24
*Foeniculum vulgare* . . . . . . . . . . . . . 30
*Garcinia atroviridis* . . . . . . . . . . . . . 11
*Garcinia schomburgkiana* . . . . . . . . 11
*Gnetum gnemon* . . . . . . . . . . . . . . . . 40
*Illicium verum* . . . . . . . . . . . . . . . . . 54
*Kaempferia galanga* . . . . . . . . . . . . .37
*Mentha arvensis* . . . . . . . . . . . . . . . . 41
*Mentha javanica* . . . . . . . . . . . . . . . 41
*Murraya koenigii* . . . . . . . . . . . . . . . .29
*Myristica fragrans* . . . . . . . . . . . . . . . 43
*Ocimum americanum* . . . . . . . . . . . 12
*Ocimum basilicum* . . . . . . . . . . . . . . 12
*Ocimum gratissimum* . . . . . . . . . . . . 12
*Ocimum tenniflorum* . . . . . . . . . . . . 12
*Pandanus amaryllifolius* . . . . . . . . . . 44
*Papaver somniferum* . . . . . . . . . . . . 48
*Phaemeria speciosa* . . . . . . . . . . . . . 55
*Pimpinella anisum* . . . . . . . . . . . . . . 30
*Piper nigrum* . . . . . . . . . . . . . . . . . . 45
*Piper sarmentosum* . . . . . . . . . . . . . 46
*Plectranthus amboinicus* . . . . . . . . . 15
*Polygonum odoratum* . . . . . . . . . . . . 47
*Sesamum indicum* . . . . . . . . . . . . . . 50
*Sinapsis alba* . . . . . . . . . . . . . . . . . . 42
*Syzygium aromatica* . . . . . . . . . . . . . 24
*Syzygium polyantha* . . . . . . . . . . . . . 49
*Tamarindus indica* . . . . . . . . . . . . . . 55
*Trigonella foenum-graecum* . . . . . . . . 31
*Zanthoxylum piperitum* . . . . . . . . . . 52
*Zingiber officinale* . . . . . . . . . . . . .33, 36

# Index by Common Name

| | |
|---|---|
| Annatto . . . . . . . . . . . . . . . . . . . . . . 10 | Pepper . . . . . . . . . . . . . . . . . . . . . . . .45 |
| *Asam gelugor* . . . . . . . . . . . . . . . . . . 11 | Pepper, wild . . . . . . . . . . . . . . . . . . .46 |
| *Asam keping* . . . . . . . . . . . . . . . . . . 11 | Polygonum . . . . . . . . . . . . . . . . . . . .47 |
| Basil . . . . . . . . . . . . . . . . . . . . . . . . . 12 | Poppy seed . . . . . . . . . . . . . . . . . . . .48 |
| Bilimbi . . . . . . . . . . . . . . . . . . . . . . . 14 | Salam leaf . . . . . . . . . . . . . . . . . . . .49 |
| Borage, Indian . . . . . . . . . . . . . . . . . 15 | Sesame seed . . . . . . . . . . . . . . . . . . .50 |
| Candlenut . . . . . . . . . . . . . . . . . . . . 16 | Shallot . . . . . . . . . . . . . . . . . . . . . . .51 |
| Cardamom . . . . . . . . . . . . . . . . . . . 17 | Sichuan pepper . . . . . . . . . . . . . . . . .52 |
| Cassia . . . . . . . . . . . . . . . . . . . . . . . 18 | Spring onion . . . . . . . . . . . . . . . . . .53 |
| Celery, Chinese . . . . . . . . . . . . . . . . 19 | Star anise . . . . . . . . . . . . . . . . . . . . .54 |
| Chilli . . . . . . . . . . . . . . . . . . . . . . . . 20 | Tamarind . . . . . . . . . . . . . . . . . . . . .55 |
| Chinese Key . . . . . . . . . . . . . . . . . . 22 | Torch ginger . . . . . . . . . . . . . . . . . . .56 |
| Cinnamon . . . . . . . . . . . . . . . . . . . . 18 | Turmeric . . . . . . . . . . . . . . . . . . . . .57 |
| Chives, Chinese . . . . . . . . . . . . . . . . 23 | Thyme, broad-leaf . . . . . . . . . . . . . . 13 |
| Chives, flat . . . . . . . . . . . . . . . . . . . 23 | |
| Cloves . . . . . . . . . . . . . . . . . . . . . . . 24 | |
| Coriander . . . . . . . . . . . . . . . . . . . . 26 | |
| Coriander, Sawtooth . . . . . . . . . . . . 25 | |
| Cumin . . . . . . . . . . . . . . . . . . . . . . 28 | |
| Curry leaf . . . . . . . . . . . . . . . . . . . . 29 | |
| Fennel . . . . . . . . . . . . . . . . . . . . . . . 30 | |
| Fenugreek . . . . . . . . . . . . . . . . . . . . 31 | |
| Five-spice powder . . . . . . . . . . . . . . 32 | |
| Galangal . . . . . . . . . . . . . . . . . . . . . 33 | |
| *Garam masala* . . . . . . . . . . . . . . . . . 34 | |
| Garlic . . . . . . . . . . . . . . . . . . . . . . . 35 | |
| Ginger . . . . . . . . . . . . . . . . . . . . . . . 36 | |
| Kencur . . . . . . . . . . . . . . . . . . . . . . 37 | |
| Lemon grass . . . . . . . . . . . . . . . . . . 38 | |
| Lime leaf, Kaffir . . . . . . . . . . . . . . . 39 | |
| Macadamia nut . . . . . . . . . . . . . . . . 16 | |
| *Melinjo* . . . . . . . . . . . . . . . . . . . . . . 40 | |
| Mint . . . . . . . . . . . . . . . . . . . . . . . . 41 | |
| Mustard seed, brown . . . . . . . . . . . 42 | |
| Nutmeg . . . . . . . . . . . . . . . . . . . . . .43 | |
| Pandan . . . . . . . . . . . . . . . . . . . . . . .44 | |